Table of Contents

1 Introduction to Virtualization 5
1.1 What is Virtualization? 8
1.2 Benefits of Virtualization 11

2 Virtual Machine Management 15
2.1 VM Creation and Configuration 18
2.2 VM Monitoring and Maintenance 20
2.3 VM Backup and Recovery 24

3 Network Virtualization 27
3.1 Virtual Switches and Networks 29
3.2 Network Virtualization Technologies 32
3.3 Network Security in Virtual Environments 35
3.4 Network Performance Optimization 38
3.5 Troubleshooting Network Issues 41

4 Storage Virtualization 45
4.1 SAN and NAS Virtualization 48
4.2 Storage Area Networks 51

5 High Availability Strategies 55
5.1 Clustering and Failover 58
5.2 Redundancy and Disaster Recovery 61

5.3 High Availability Best Practices 63

6 Cost Optimization Techniques 67
6.1 Resource Consolidation 69
6.2 Right-Sizing VMs 72
6.3 Energy Efficiency 75
6.4 Licensing and Cost Management 78
6.5 Automation and Efficiency 81

7 Security in Virtualized Environments 85
7.1 Securing Virtual Machines 88
7.2 Security Best Practices 91

1 Introduction to Virtualization

Virtualization fundamentally alters how we approach computing. Instead of dedicating physical hardware to single applications, virtualization allows multiple virtual machines (VMs) to run concurrently on a single physical host. Think of it as having several apartments within a single building; each apartment (VM) operates independently, unaware of its neighbours, yet all share the building's (physical host's) resources. This seemingly simple concept has revolutionized IT infrastructure, offering unparalleled flexibility and efficiency. The advantages are numerous and impactful. Perhaps the most obvious benefit is cost reduction. Consolidating numerous physical servers into fewer, more powerful virtualized hosts significantly reduces hardware expenditures, including power consumption and physical space requirements. This translates to substantial long-term savings for organizations of all sizes. Furthermore, virtualization enhances resource utilization. By dynamically allocating resources to VMs as needed, you avoid the waste associated with underutilized physical servers. A VM can be quickly scaled up or down, adapting seamlessly to fluctuating demands.

This flexibility extends to software deployments. Testing new applications or operating systems becomes significantly easier and less risky. You can create VMs to test software in isolated environments, preventing conflicts with existing systems and avoiding potential disruptions to production environments. Disaster recovery is greatly simplified too; creating and restoring VMs from backups is significantly faster and less complex than managing physical servers. This agility reduces downtime and accelerates recovery time in the event of a failure.

Let's delve into the practical aspects of creating and managing VMs. Most virtualization platforms offer intuitive interfaces for this purpose. Typically, you start by specifying the VM's operating system, allocating resources like CPU cores, RAM, and disk space, and then configuring network settings. Careful resource allocation is key; over-provisioning wastes resources, while under-provisioning can lead to performance bottlenecks. Finding the right balance is a crucial skill.

Post-creation, monitoring is vital. Regular monitoring helps detect issues before they escalate. Tools are readily available to track VM performance metrics like CPU utilization, memory usage, and disk I/O. These tools often provide alerts for unusual activity or potential problems. Proactive monitoring allows for timely intervention, preventing performance degradation and ensuring continued stability. Equally crucial is routine maintenance. This might involve updating the guest operating system, applying security patches, or optimizing

VM settings to enhance performance.

Protecting your VMs against data loss is paramount. Implement a robust backup and recovery strategy. Regular backups, ideally using an automated system, are critical. Testing the recovery process is equally important; ensure your backup system works effectively and that you can restore VMs quickly and efficiently. Without a proper backup strategy, a VM failure can translate to significant data loss and considerable downtime. The time investment is well worth it for the peace of mind and resilience it provides.

Consider the specific requirements of your applications and workloads when designing your virtual infrastructure. Different applications have varying resource needs. A database server, for instance, demands significantly more resources than a simple web server. Right-sizing VMs ensures that each application receives the optimal resources it needs, improving performance and resource utilization without unnecessary over-provisioning. This optimization is essential for maximizing cost efficiency and operational effectiveness. Understanding your workload profiles is crucial for effective virtualization planning.

Virtualization opens up a world of possibilities, from streamlined software development and testing to improved disaster recovery and enhanced cost efficiency. The benefits are considerable, justifying the investment in understanding and implementing virtualization strategies. With careful planning and diligent management, your organization can harness the full power of virtualization to optimize its IT infrastructure, increase operational agility, and achieve significant long-term cost savings. Embrace the potential and embark on a journey towards a more efficient and resilient IT landscape.

1.1 What is Virtualization?

Virtualization fundamentally abstracts physical hardware resources, creating virtual versions that behave like independent physical entities. Imagine a single, powerful server acting as a host for numerous virtual servers, each with its own operating system, applications, and resources. This allows for efficient resource utilization and simplifies management compared to managing numerous physical machines. It's a bit like having a single apartment building, instead of many individual houses, each with its own utilities and maintenance needs. The building manager (the virtualization system) handles the overall infrastructure, while each apartment (the virtual machine) functions autonomously.

This abstraction offers several key advantages. Firstly, it dramatically improves resource utilization. Instead of having many underutilized physical servers sitting idle, virtualization allows you to consolidate workloads onto fewer physical machines, maximizing their potential. Secondly, virtualization simplifies management. Instead of needing to physically manage and maintain dozens of servers, administrators can manage everything from a central console, significantly reducing administrative overhead. Thirdly, it enhances flexibility. Spinning up or shutting down a virtual machine is a matter of minutes, as opposed to the hours or days it could take with physical servers, allowing for faster deployment of applications and faster response to changing business needs.

Let's consider a practical example: a small business with limited IT resources. Using virtualization, they could run their web server, database server, and email server all on a single powerful physical server. This simplifies management, reduces energy costs, and minimizes the space required for hardware. The same concept scales to larger organizations; they can consolidate thousands of servers into a smaller number of physical hosts, significantly reducing capital expenditure and operational costs.

A key component in understanding virtualization is the concept of a hypervisor. This is the software that sits between the physical hardware and the virtual machines. It's responsible for allocating resources, managing the virtual machines, and providing the virtualization layer. Think of it as the building's foundation and central utilities— providing the essential services that each apartment (virtual machine) relies upon. There are two main types of hypervisors: Type 1, also known as bare-metal hypervisors, directly interact with the hardware, while Type 2, or hosted hypervisors, run on top of an existing operating system. Choosing between these depends on your specific needs and infrastructure.

One of the significant challenges with virtualization is the potential for performance degradation if not properly managed. Over-provisioning virtual machines can lead to contention for resources, resulting in performance bottlenecks and application slowdowns. This is why careful planning and resource allocation are crucial, involving thorough assessment of application requirements, and ensuring sufficient resources are dedicated to each virtual machine. Regular monitoring of CPU, memory, and disk I/O is also essential to identify any potential issues before they impact performance. Right-sizing VMs is key; start small, monitor performance, and only allocate more resources when genuinely needed.

Security in a virtualized environment is also a critical concern. While virtualization itself doesn't introduce new security risks, it does change the security landscape. Securing virtual machines requires a multi-layered approach, including implementing strong passwords, regularly patching operating systems and applications, using firewalls, and employing intrusion detection systems. Network segmentation within the virtual environment can further isolate virtual machines and protect against lateral movement of threats. Regular security audits and vulnerability scans are also essential to maintain a secure virtualized infrastructure. Moreover, implementing robust backup and recovery procedures is a crucial aspect of securing the virtual environment. Regular backups of virtual machines are essential to ensure business continuity in the event of hardware failure or data loss. These backups should be stored securely and tested regularly to ensure recoverability.

Understanding virtualization requires grasping its core concepts. It's not just about creating virtual machines; it's about efficiently managing resources, simplifying administration, and enhancing flexibility and security. By carefully considering your needs and implementing appropriate strategies, you can leverage the many benefits of virtualization while mitigating the potential challenges. The ultimate goal is to build a highly available, cost-effective, and secure virtualized infrastructure that supports your business needs. This involves ongoing monitoring, adaptation, and a commitment to best practices.

1.2 Benefits of Virtualization

Reduced hardware costs are a significant advantage. Instead of purchasing multiple physical servers, a single physical server can host numerous virtual machines, leading to substantial savings on hardware acquisition, power consumption, and physical space. This translates directly to a lower total cost of ownership (TCO) over time, a crucial factor for any business, particularly startups and small to medium-sized enterprises (SMEs). Furthermore, this consolidation reduces the complexity of managing numerous individual servers, freeing up IT staff to focus on other strategic initiatives. The potential for capital expenditure savings is amplified when considering the reduced need for server rooms, cooling systems, and associated infrastructure.

Improved resource utilization is another key benefit. Virtualization allows for dynamic allocation of resources like CPU, memory, and storage, to virtual machines based on their needs. This eliminates resource wastage often seen in underutilized physical servers. For example, a virtual machine running a database application can be dynamically allocated more resources during peak hours and scaled back when demand lessens. This flexibility ensures optimal performance across all virtual machines, leading to better utilization of the overall infrastructure and avoiding the common issue of "over-provisioning" physical resources to ensure sufficient capacity. This efficient resource allocation is further enhanced through the capability of live migration, enabling the seamless movement of virtual machines between physical hosts with minimal disruption to running applications.

Increased agility and flexibility are paramount in today's dynamic business environment. Virtualization facilitates rapid deployment and scaling of applications and infrastructure, enabling businesses to respond quickly to changing market demands. Spinning up a new virtual machine takes mere minutes, compared to the days or weeks required for provisioning a physical server. This speed allows for faster testing, development, and deployment cycles, fostering innovation and reducing time-to-market. Businesses can easily test new applications or scale their infrastructure up or down according to demand, without the need for significant capital investment in new hardware. This flexibility is particularly advantageous for organizations dealing with fluctuating workloads or seasonal peaks in demand.

Enhanced disaster recovery and business continuity capabilities are made possible with virtualization. Creating backups and replicas of virtual machines is significantly simpler and faster compared to physical servers. This ease of replication allows for the quick recovery of services in case of hardware failure or other unforeseen circumstances. Furthermore, virtual machines can be easily migrated to alternative data centers in case of a disaster, ensuring business continuity with minimal downtime. High availability features such as clustering and failover mechanisms are easily implemented, further improving resilience. The ability to quickly spin up replacement virtual machines from backups minimizes recovery time objectives (RTO) and recovery point objectives (RPO), significantly reducing the impact of disruptions on business operations.

Improved application availability contributes significantly to increased overall productivity. The rapid provisioning and deployment capabilities of virtual machines allow for quicker recovery times from planned and unplanned outages. This enhanced uptime ensures continuous access to critical applications and data, minimizing business disruption and enhancing overall productivity. Additionally, the ability to easily migrate virtual machines between physical servers enables scheduled maintenance without affecting service availability. For example, a planned hardware upgrade can be executed seamlessly without any impact on the applications running on the VMs; this planned maintenance contributes to long-term system stability, further boosting application availability. This proactive approach to maintenance minimizes disruption and ensures a higher level of system reliability.

Simplified management reduces IT overhead. The centralized management capabilities of virtualization platforms streamline the administration of multiple virtual machines, servers, and storage, simplifying and improving IT operational efficiency. This consolidation reduces the complexities associated with managing diverse physical hardware, allowing IT staff to focus on more strategic initiatives. Features such as automated provisioning and management tools significantly improve efficiency and reduce administrative burden. The ability to manage resources and perform tasks from a single console further improves the overall manageability of the virtualized infrastructure, making IT operations more efficient and less complex. This ultimately frees up resources and allows for better utilization of IT staff's time and expertise.

2 Virtual Machine Management

Creating a new virtual machine (VM) involves several steps. First, you'll need to choose the appropriate hypervisor, considering factors like compatibility with your existing hardware and operating system, and the features you require. Next, allocate sufficient resources—CPU cores, RAM, and storage—to the VM, taking into account the workload's demands. Insufficient resources can lead to performance bottlenecks, whereas over-allocation is wasteful. Careful planning is key. Remember to select the desired operating system for your VM. The installation process will vary slightly depending on the hypervisor and operating system, but generally involves mounting an ISO image or pointing to a network installation source. Post-installation, configure the VM's networking settings, ensuring proper connectivity to your network infrastructure. This may involve assigning a static IP address or using DHCP. Finally, install any necessary applications and drivers to make the VM fully functional, always remembering security best practices throughout the process. Regularly review resource allocation to optimize performance and cost efficiency. Proactive management prevents unexpected issues.

Monitoring your VMs is crucial for maintaining their health and performance. Utilize your hypervisor's built-in monitoring tools to track key metrics such as CPU utilization, memory consumption, disk I/O, and network traffic. Identify and address performance bottlenecks proactively. Establishing performance baselines allows you to easily spot deviations indicating potential issues. Regularly review logs for errors or warnings. These often provide valuable insights into system problems. Consider implementing automated alerts for critical events, enabling swift intervention and preventing larger problems. Proactive monitoring reduces downtime and maintains a stable virtual infrastructure. Remember, understanding your system's normal behavior is the foundation of effective problem-solving.

Regular maintenance is essential for VM longevity. This involves updating the guest operating systems and applications to the latest versions, patching security vulnerabilities, and performing regular software updates. These updates frequently include performance enhancements and crucial security fixes. Defragmenting virtual disks can improve performance and prevent fragmentation-related issues. This is especially relevant for VMs with heavy write operations. Regularly check disk space utilization and delete unnecessary files to free up storage. Remember, proactive maintenance minimizes the risk of system failures and data loss. This includes implementing a robust backup and recovery strategy as described in the next section. Consider scripting or automating routine tasks for efficiency and consistency. This saves time and minimizes errors in your maintenance procedures.

Backing up your VMs is a critical aspect of ensuring business continuity and data protection. The best strategy involves using a combination of methods, such as regular snapshots, full backups, and incremental backups. These methods offer different recovery time objectives (RTOs) and recovery point objectives (RPOs). Snapshots, frequently used for rapid recovery, provide a point-in-time image of the VM's state. Full backups provide a complete copy of the VM's data, ideal for archival and disaster recovery. Incremental backups capture only changes since the last full or incremental backup, optimizing storage usage. The frequency of backups depends on your data change rate and recovery needs. Consider offsite backup storage to protect against physical loss. Test your backup and recovery strategy regularly to ensure its effectiveness and make any needed adjustments. Regular testing ensures that the backup and recovery process works and can be performed within an acceptable time frame.

Recovery from VM failures requires a well-defined process. Start by identifying the cause of the failure; this is crucial for preventing future occurrences. Then, use your backups to restore the VM to a functional state. Depending on your backup method and chosen RTO/RPO, this could involve restoring from a full backup, a snapshot, or applying incremental backups. After restoring, validate the VM's functionality by verifying data integrity and application performance. It's also essential to investigate the root cause of the failure and implement preventative measures. Document the entire recovery process, including the time it took, the challenges faced, and any lessons learned. This documentation is crucial for improving future responses and disaster recovery plans. Your recovery procedure should be thoroughly tested and documented for efficient and effective responses to downtime.

2.1 VM Creation and Configuration

Creating a new virtual machine (VM) begins with selecting the appropriate hypervisor. Consider your specific needs; for instance, VMware vSphere offers robust features and extensive management capabilities, ideal for larger enterprise environments. Alternatively, Hyper-V, integrated into Windows Server, presents a simpler, more cost-effective solution suitable for smaller deployments. Careful consideration of your budget and technical expertise will guide your choice. Remember, the hypervisor forms the bedrock of your entire virtualized infrastructure; choosing wisely ensures a stable and efficient foundation. This decision will also influence the tools and techniques you'll employ throughout the entire VM lifecycle.

The process itself varies slightly based on your chosen hypervisor, but the fundamental steps remain consistent. You'll generally begin by specifying the VM's resource allocation. This includes crucial parameters like the amount of RAM, CPU cores, and disk space. Over-provisioning resources can lead to wasted expenditure; under-provisioning results in performance bottlenecks. Thorough planning and accurate estimations of resource requirements are critical. Remember to account for future growth and potential application demands. Analyze your workload characteristics to optimize resource allocation. A thorough understanding of your applications' resource needs is essential for efficient resource allocation. For example, a database server will require significantly more RAM and storage than a simple web server.

Next, you'll need to select the operating system you wish to install within the VM. This choice depends entirely on the applications you plan to run. The operating system must be compatible with both the hypervisor and the intended software. Consider licensing costs when making your choice; some operating systems are more expensive than others. Ensure you have the correct installation media ready and accessible to the hypervisor. Also, remember to adhere to licensing agreements; unauthorized usage can lead to legal ramifications. Furthermore, factor in the potential security implications of each OS; some are inherently more secure than others.

Once the OS is installed, the configuration process begins. This is where you install the necessary drivers and software needed for your applications to run smoothly. This stage demands meticulous attention to detail. Correctly configuring network settings is crucial; ensure the VM has appropriate access to your network infrastructure. This frequently involves configuring network adapters and assigning IP addresses within the virtual network. Proper networking is vital to ensure the VM can communicate with other systems

and the internet. Misconfigurations in this phase can cause connectivity problems and limit functionality.
Security is paramount. Implementing robust security measures from the outset protects your VM from external threats and potential vulnerabilities. Consider employing firewalls, intrusion detection systems, and regular security updates to ensure the integrity of your virtualized environment. Strong passwords and access control lists (ACLs) further enhance security. Remember that vulnerabilities within a VM can compromise the entire system, especially in poorly planned network configurations. The application of patches, updates, and security best practices should be prioritized.
Finally, remember to thoroughly document every step of the process. This documentation, including detailed specifications of hardware and software configurations, is essential for troubleshooting and maintenance. It's also beneficial if you need to recreate the VM later. Comprehensive and accurate documentation saves valuable time and effort during future maintenance or troubleshooting activities. Think of it as insurance; should something go wrong, detailed notes will significantly reduce downtime and remediation efforts. This also enables others to manage the VM effectively if necessary. A well-maintained logbook is an invaluable asset.

2.2 VM Monitoring and Maintenance

Regular monitoring is the cornerstone of efficient VM management. Think of your VMs as vital organs within a digital body; continuous observation is essential for early detection of problems. Employing readily available monitoring tools, both built into your hypervisor and third-party options, allows you to track key performance indicators (KPIs) such as CPU utilization, memory consumption, disk I/O, and network bandwidth. Setting up alerts for critical thresholds—like CPU consistently exceeding 80%—ensures prompt intervention before performance degradation impacts applications or users. This proactive approach is far more efficient than reactive troubleshooting. Don't underestimate the value of simple, readily available metrics. They can reveal hidden inefficiencies, ultimately saving time and preventing larger issues down the line. Understanding these baseline metrics forms the foundation of effective VM maintenance.

Regular patching and updates are crucial for maintaining the security and stability of your VMs. Operating system vulnerabilities can be exploited by malicious actors, potentially leading to data breaches or system failures. Schedule regular patching cycles, leveraging your hypervisor's capabilities to deploy patches efficiently and with minimal disruption. Prioritize critical security patches and test them in a non-production environment before rolling them out to production VMs. Consider using automated patch management tools to streamline the process and eliminate the risk of human error. Thorough testing is paramount to ensure compatibility with your applications and prevent unexpected downtime. Regular updates also include keeping your hypervisor, virtualization management software, and associated drivers updated, enhancing overall system security and resilience.

Resource allocation is a critical aspect of VM maintenance. Over-provisioning resources can lead to unnecessary expenses, while under-provisioning can result in performance bottlenecks. Regularly review VM resource usage to identify any inconsistencies or inefficiencies. Tools and features provided by most hypervisors allow you to easily monitor resource consumption over time. This analysis will help you optimize resource allocation, potentially reducing operating costs and improving performance. Right-sizing VMs - adjusting their resource allocation to match their actual needs - is a significant cost-saving strategy that also improves overall system stability. Careful observation of resource trends allows you to anticipate future demands and make proactive adjustments. Avoid the trap of "just in case" over-provisioning; it is counterproductive.

Proactive log analysis is invaluable. Examine system and application logs regularly to identify potential issues before they escalate into major problems. Modern hypervisors offer centralized log management capabilities. These systems help you to track events, troubleshoot errors, and identify potential security risks. Setting up log monitoring tools with automated alerting can substantially reduce the mean time to resolution (MTTR) for various issues. This system ensures that you are aware of minor problems before they develop into major outages. Regular review helps improve system performance and strengthens overall security. Prioritize analyzing logs from critical system components and applications first.

Regular backups are the ultimate safety net for your virtual environment. Implementing a robust backup and recovery strategy is essential to minimize data loss and downtime in case of unexpected events such as hardware failures, software glitches, or cyberattacks. Consider using a combination of different backup methods such as full backups, incremental backups, and differential backups. A well-defined backup schedule will minimize the risk of significant data loss in case of a disaster. Regularly test your backup and recovery procedures to ensure they function as expected. This verification prevents unexpected issues during actual recovery scenarios. Remember, a backup plan is only as good as its regular testing and implementation. Failure to test the procedure negates the whole process. A comprehensive backup strategy should also incorporate offsite backups to protect against physical damage or site-wide disasters.

Performance testing should be a regular part of your VM maintenance routine. Conduct periodic performance testing to identify potential bottlenecks and ensure your VMs meet the performance requirements of their applications. Tools that simulate real-world workloads are key for identifying inefficiencies that might go unnoticed under normal operation. Performance tests allow for early detection of potential problems before they impact your end users. This proactive approach is crucial for maintaining a high level of service availability. Regular testing also helps optimize resource allocation, ensuring the correct balance between performance and cost efficiency. Consider using various testing methodologies, such as stress testing and load testing, to thoroughly evaluate your VMs' performance under different conditions.

2.3 VM Backup and Recovery

Regular backups are your lifeline in the volatile world of virtual machines. Without them, a single hardware failure, a rogue script, or even a simple misconfiguration can lead to significant data loss, potentially crippling your operations. Consider this your wake-up call to prioritize a robust VM backup and recovery strategy. Implementing one is not just good practice; it's essential for business continuity.

Choosing the right backup method is crucial, and the ideal solution often depends on your specific needs and resources. Full backups create a complete copy of your VMs, offering a straightforward recovery process but requiring more storage space and backup time. Incremental backups, on the other hand, only capture changes made since the last backup, significantly reducing storage needs and backup times. Differential backups represent a compromise, storing only the differences since the last full backup, striking a balance between speed and storage consumption. The best approach may involve a combination of these techniques - a full backup followed by a series of incremental or differential backups, ensuring rapid recovery while managing storage efficiently.

Your storage strategy is equally vital. Local backups, stored directly on the host server, are simple to implement but vulnerable to the same failures that could impact your VMs. Network-attached storage (NAS) or storage area networks (SANs) offer redundancy and protection against host failures, greatly enhancing data security. Cloud-based solutions, such as those offered by major cloud providers, provide an additional layer of disaster recovery, protecting your data from even catastrophic events such as fire or natural disasters. Remember to factor in costs, storage capacity, and bandwidth requirements when making your choice. Don't just choose the cheapest option; select the option that offers the best balance between cost and resilience.

Beyond the 'what', the 'how' is just as important. Using specialized backup software designed for virtual environments streamlines the entire process. Many solutions automate scheduling, provide granular control over what data is backed up, and even offer features like built-in compression and deduplication to optimize storage utilization. These features can be invaluable, saving time and resources. Familiarize yourself with your chosen software's features, ensuring you understand how to configure schedules, manage backups, and perform restorations. This proactive approach drastically reduces recovery time in the event of an incident.

Testing your backup and recovery plan is not an optional step; it's a necessity. Regularly perform test restores to ensure your backups are valid and your recovery procedures work flawlessly. This preventative measure uncovers potential issues and allows you to refine your process before an actual disaster strikes. Nothing is more frustrating than discovering a critical flaw in your plan when you desperately need it. Don't wait for the crisis; plan for it now. Simulate a scenario—perhaps a VM crash or a complete host failure—to identify bottlenecks and areas for improvement. Document your procedures meticulously, creating a step-by-step guide that anyone on your team can follow easily, regardless of their expertise level. This is crucial for ensuring a smooth and efficient recovery process in a high-pressure situation. Remember, a well-documented plan is a successful plan.

Finally, consider the human element. Train your team on backup and recovery procedures. Establish clear roles and responsibilities to ensure everyone knows what to do in the event of an emergency. Regular training sessions, complemented by readily available documentation and step-by-step guides, empower your team to handle incidents effectively, minimizing downtime and ensuring business continuity. A well-trained team is your best defense against unexpected data loss. A well-rehearsed emergency response plan reduces stress and ensures a swift, competent recovery. Never underestimate the power of preparation and proactive training. The time you invest in these measures will pay off handsomely in the long run, minimizing disruption and preserving your valuable data. Remember, investing in a robust backup and recovery strategy isn't just about protecting your data; it's about safeguarding your business.

3 Network Virtualization

Virtual switches form the bedrock of network virtualization. They're software-defined networks, allowing you to create logical networks independent of the underlying physical hardware. Think of them as virtual routers and switches, managing network traffic within your virtual environment. This flexibility allows for isolating virtual machines, creating VLANs for segmented security and improved network management. Proper configuration is key; ensuring adequate bandwidth allocation and implementing Quality of Service (QoS) policies to prioritize critical applications. Don't underestimate the power of efficient switch configuration—it significantly impacts overall network performance. Consider your needs carefully and choose a virtual switch that aligns with your virtualization platform, offering features like port mirroring and advanced security protocols for robust network management.

Understanding the different Network Virtualization technologies available is critical. Software Defined Networking (SDN) offers central control over your network, enabling dynamic resource allocation and simplified management. Virtual LANs (VLANs) provide logical network segmentation, enhancing security by isolating different groups of virtual machines. VXLAN, an overlay network technology, extends VLAN capabilities across multiple physical networks, especially useful in data centers with multiple locations. These technologies, when used correctly, provide significant advantages in flexibility and scalability compared to traditional networks, allowing for rapid deployment of new virtual machines and efficient resource utilization. The key is choosing the right technology based on your specific requirements and understanding its limitations.

Security in a virtualized network requires a multi-layered approach. Firewall configuration at both the virtual switch and individual VM levels is crucial. Implement intrusion detection and prevention systems (IDS/IPS) to monitor network traffic for malicious activity. Regular security audits and vulnerability scans are essential, keeping your virtualized network up-to-date with the latest security patches and software updates. Access control is paramount; restricting network access to authorized users and machines only. Remember, a robust security posture isn't a one-time task but an ongoing process of proactive measures to maintain a secure virtual environment. Neglecting security is a significant risk, leaving your sensitive data vulnerable to breaches.

Optimizing network performance is a balancing act. Effective bandwidth allocation, as previously mentioned, is crucial to prevent bottlenecks and ensure applications run smoothly. Properly sizing your virtual network infrastructure is essential to accommodate current and future needs. Network monitoring tools provide real-time visibility into network performance, identifying potential issues before they impact applications or users. Regular performance testing, under various load conditions, will help you proactively optimize your network configuration for peak performance. Monitoring allows you to identify network bottlenecks, optimize traffic flow, and address problems before they cause serious disruptions to your operations, ultimately ensuring high availability and user satisfaction.

Troubleshooting network issues in a virtualized environment can be complex. Start by isolating the problem—is it a virtual machine, the virtual switch, or the physical network infrastructure? Use network monitoring tools to analyze traffic patterns and pinpoint the source of the issue. Detailed logs from virtual switches and VMs can offer crucial clues. Common issues include network connectivity problems, DNS resolution failures, and bandwidth limitations. Systematic troubleshooting, combined with a good understanding of your virtual network architecture, will help you efficiently identify and resolve these issues, minimizing downtime and maintaining a stable network. Remember to document your troubleshooting steps for future reference, and consider creating a troubleshooting checklist to streamline the process.

3.1 Virtual Switches and Networks

Virtual switches are the cornerstone of virtual networking, acting as the central point of communication within a virtualized environment.

They allow virtual machines (VMs) to communicate with each other and with the physical network, all without the need for dedicated physical network hardware for each VM. Think of them as software-defined routers, directing network traffic based on pre-configured rules. Efficiently managing these switches is crucial for optimal performance and network security.

Understanding the different types of virtual switches is essential. Some offer basic functionality, simply bridging connections between VMs. Others offer more advanced capabilities, including VLAN support, traffic shaping, and quality of service (QoS) features. Choosing the right type depends on the specific requirements of your virtual environment. For instance, if you have a high-bandwidth application requiring prioritization, a switch with QoS would be essential. Conversely, a simpler switch suffices for environments with less stringent network needs.

Creating and configuring virtual switches involves several steps, starting with the selection of the appropriate type, followed by its assignment to the virtual network and VMs. This process involves specifying parameters such as the virtual network's IP address range, subnet mask, and gateway. Proper configuration is crucial to avoid network conflicts and ensure smooth communication. For example, improperly assigned IP addresses lead to network connectivity issues, which can significantly impact productivity. Therefore, meticulous attention to detail during configuration is paramount.

Security considerations are paramount when dealing with virtual switches. They often become a prime target for malicious activity, potentially affecting the entire virtual environment. Implementing security measures like access control lists (ACLs), port security, and network segmentation becomes critical. ACLs, for example, allow you to define rules that control which VMs can communicate with each other based on specific criteria, effectively creating a barrier against unauthorized access. This approach forms a robust defense against network-based attacks.

Monitoring the performance of virtual switches is an ongoing task. Regularly checking metrics like bandwidth utilization, latency, and packet loss is crucial for identifying potential bottlenecks and addressing them proactively. Tools such as VMware vCenter or Hyper-V Manager provide valuable insights into switch performance, alerting you to any issues that might require attention. Ignoring these crucial indicators can lead to degraded network performance impacting the entire virtualized infrastructure. You should always be proactive in identifying and addressing such issues.

Troubleshooting network issues within a virtualized environment can be challenging. However, a systematic approach is key to rapid resolution. This often starts with verifying the basic network configurations of the VMs and the switches, then checks for physical network connectivity. Leveraging network monitoring tools helps narrow down the problem, pinpointing the root cause of the issue. Remember that a thorough understanding of the virtual network's architecture and components is essential for effective troubleshooting. For instance, identifying the source of a connectivity problem requires understanding how the virtual switches connect VMs and how they interconnect with the physical network.

Network virtualization provides flexibility and scalability, enabling efficient resource allocation and enhanced network security. However, the effectiveness of virtualization relies heavily on the proper design, configuration, and maintenance of virtual switches. Remember that a well-planned virtual network architecture and secure configuration practices are vital for maintaining a stable, high-performing, and secure virtualized environment. Therefore, paying careful attention to every aspect of virtual switch management ensures optimized network performance and minimizes the risk of disruptions or security breaches. Thorough planning and attention to detail are essential components of success in this domain.

3.2 Network Virtualization Technologies

Network virtualization decouples the network from the physical hardware, allowing for flexible and efficient network management. Think of it like this: instead of having dedicated physical network switches and routers for each server, you create a virtual network overlay on top of your existing infrastructure. This allows for dynamic allocation of network resources, simplified management, and increased scalability. This is crucial in today's dynamic IT landscape. Implementing network virtualization often involves using software-defined networking (SDN) technologies. SDN separates the network control plane (the brains that make decisions about network traffic) from the data plane (the actual forwarding of data packets). This separation offers greater control and programmability. Imagine having a central dashboard to manage your entire virtual network, rather than dealing with individual physical devices - a significant improvement in efficiency, especially as your network grows. This allows for easier automation and deployment of new network services, ultimately enhancing operational efficiency.

Virtual Local Area Networks (VLANs) are a common networking technology that becomes even more powerful within a virtualized environment. VLANs allow you to segment your network logically, even if the physical infrastructure remains unchanged. For example, you could create separate VLANs for different departments or applications, improving security and isolating network issues. Careful planning and implementation of VLANs are vital for network segmentation and security. Without them, the virtualized environment's flexibility would be dramatically reduced.

VXLAN (Virtual Extensible LAN) offers another layer of abstraction. VXLAN extends VLAN functionality to handle larger networks and offers better scalability. It encapsulates VLAN traffic in UDP packets, allowing for greater flexibility in how you structure your virtual network. Unlike traditional VLANs that are limited by the number of VLAN IDs, VXLAN offers a much larger address space and supports virtual networks across multiple physical locations. This is especially beneficial in large data centers or cloud environments where scalability and flexibility are paramount.

Consider the use of virtual switches. These software-based switches replicate the functionality of physical switches but offer greater flexibility and manageability. You can easily create, delete, and modify virtual switches within your virtual environment. This enables quicker setup, deployment of services and resources, as well as reduced administrative overhead. By efficiently managing your virtual switches, you ensure optimal performance.

Network virtualization offers significant advantages in terms of cost savings. By consolidating network resources and simplifying management, you reduce the need for expensive physical hardware and specialized personnel. This translates to lower capital expenditures (CAPEX) and operating expenses (OPEX). This efficient use of resources can dramatically decrease the overall cost of your infrastructure, a crucial benefit in any budget-conscious environment.

Security is paramount in any network environment, and this is particularly true in a virtualized setting. Network virtualization technologies, when implemented correctly, can actually enhance security. For example, VLANs and VXLAN allow for finer-grained control over network access, effectively isolating sensitive data and applications. This enhances the effectiveness of your security policies and reduces the potential impact of security breaches.

Troubleshooting network issues in a virtualized environment might initially seem more complex. However, the centralized management provided by SDN and virtual switches often simplifies this process.

Centralized logging and monitoring tools provide a comprehensive view of your network, making it easier to identify and resolve problems quickly. Proactive monitoring and the use of readily available troubleshooting guides and tools are key elements for keeping your virtual network running efficiently.

Effective network virtualization requires careful planning and implementation. Understanding the different technologies available and choosing the right tools for your specific needs is vital. Consider factors like scalability, security, and manageability. Don't be afraid to consult with experienced network professionals to get help or advice as you implement these technologies. Starting with a well-defined strategy ensures a smoother transition to a virtualized network. Remember, a well-planned approach mitigates risks and ensures optimal results. This initial planning stage is essential for long-term success.

3.3 Network Security in Virtual Environments

Securing your virtualized network isn't just about adding firewalls; it's about a layered approach, a holistic strategy. Think of it like building a castle: you need strong walls, but also moats, guards, and well-placed archers. First, you must segment your network. Don't leave all your VMs in one big, vulnerable space. Use VLANs (Virtual Local Area Networks) to separate sensitive data from less critical applications. Imagine your accounting department's servers separated from your marketing team's—a crucial distinction for data privacy and security. VLANs act as virtual walls, controlling traffic flow and preventing unauthorized access.

Implementing robust access controls is another critical step. Utilize strong authentication methods—think beyond basic passwords. Multi-factor authentication, incorporating something you know, something you have, and something you are, significantly enhances security. This layered approach makes it exponentially more difficult for malicious actors to breach your systems. Regularly review and update access permissions, removing any unnecessary privileges. This proactive approach minimizes potential vulnerabilities.

Intrusion detection and prevention systems (IDPS) are your virtual guards. They continuously monitor network traffic for malicious activities, alerting you to potential threats and automatically blocking suspicious behavior. Deploy these systems at strategic points within your virtual network—think of them as strategically positioned archers, ever vigilant against potential attacks. Regularly update the signature databases of your IDPS to ensure that they can identify and defend against the latest threats. Failing to do so is like leaving your archers with outdated weapons.

Regular patching and updating of your virtual infrastructure, including the hypervisor itself and guest operating systems, is non-negotiable. Software vulnerabilities are constantly being discovered, and promptly applying patches minimizes your exposure to exploits. Schedule regular maintenance windows for this crucial task, ensuring it's not overlooked amidst other priorities. This proactive approach is essential—ignoring updates leaves your virtual castle with gaping holes in its walls.

Beyond the technical measures, strong security policies and procedures are paramount. Establish clear guidelines for acceptable use, data handling, and incident response. These policies should be regularly reviewed and updated. Employee training is equally crucial; your staff needs to understand and follow these policies. A well-trained workforce is your castle's most valuable asset. Regular security awareness training, including phishing simulations, should be incorporated into your overall security strategy. Don't underestimate the power of education—well-trained personnel can often identify and prevent threats before they escalate.

Virtualized environments, while offering significant benefits, also present unique security challenges. The ease of cloning and moving VMs can be exploited if proper safeguards are not in place. Secure your VM templates and ensure that only authorized personnel can create and modify them. Consider using a centralized VM management solution that enables robust access control and auditing capabilities. This approach provides granular control and a comprehensive audit trail, enhancing both security and accountability.

Regular security audits are vital. Independent assessments of your virtualized network, performed by qualified professionals, identify vulnerabilities you may have overlooked. They offer an objective, external perspective on your security posture. These audits should be scheduled regularly, and remediation of identified issues should be prioritized. This ongoing process ensures that your virtual castle remains strong and secure against ever-evolving threats. Don't wait for a breach to discover weaknesses; actively identify and mitigate vulnerabilities.

Remember, security in a virtual environment is an ongoing process, not a one-time event. It requires constant vigilance, adaptation, and a commitment to best practices. Treat your virtual network with the same level of security that you would accord to your physical infrastructure—or even more so. The agility and flexibility of virtualization can be a double-edged sword if not handled carefully. By implementing a multi-layered security strategy and maintaining a proactive approach, you can effectively protect your valuable data and ensure the continued operation of your business. Invest in robust security measures; the cost of a breach far outweighs the investment in prevention.

3.4 Network Performance Optimization

Efficient network performance is paramount in a virtualized environment. Slowdowns and latency directly impact application responsiveness and user experience, potentially leading to lost productivity and revenue. Optimizing your network is therefore not just about speed; it's about ensuring the stability and reliability of your entire virtual infrastructure. This requires a multifaceted approach.

Begin by carefully assessing your current network infrastructure. Identify any bottlenecks - overloaded switches, saturated links, or poorly configured virtual networks. Tools like network monitoring software, capable of providing real-time visibility into traffic patterns and resource utilization, are invaluable here. Detailed analysis will pinpoint areas needing immediate attention, such as specific virtual machines consistently consuming excessive bandwidth. Don't underestimate the power of thorough, data-driven diagnosis.

Next, consider the network architecture itself. Are your virtual machines effectively segmented? Employing VLANs (Virtual Local Area Networks) can significantly improve network security and performance by isolating different groups of VMs. Furthermore, VXLAN (Virtual Extensible LAN) provides scalability across multiple physical networks, allowing for flexible network segmentation and management even in large-scale deployments. Efficiently utilizing these technologies is crucial for better resource allocation and prevention of broadcast storms.

The sizing of your virtual switches is another critical factor. Oversized switches may seem like an advantage, but they can lead to unnecessary resource consumption. Conversely, undersized switches become bottlenecks, impacting performance considerably. Properly sizing your virtual switches, based on projected traffic and future growth, involves careful planning and potentially iterative adjustments as your needs evolve. Aim for optimal balance, avoiding both over-provisioning and under-provisioning.

Network configuration plays a crucial role. Incorrect settings, like poorly chosen MTU (Maximum Transmission Unit) sizes or mismatched network configurations between physical and virtual environments, can severely hinder performance. Regular checks and validation of all network settings, ensuring consistency across your virtual infrastructure, are essential. This includes verifying proper routing protocols and firewall rules, eliminating any conflicts or unnecessary restrictions on data flow.

Resource allocation within your virtual environment is often overlooked. Over-allocating resources to specific VMs impacts the overall network, whereas under-allocation leads to performance degradation. Monitoring resource utilization, especially bandwidth and CPU, is crucial for efficient allocation. Adjusting resource allocations dynamically, based on real-time performance data, allows for optimal network utilization and prevents bottlenecks from developing.

Finally, consider implementing QoS (Quality of Service) policies. QoS allows you to prioritize specific types of traffic, such as critical application data, ensuring that these always have sufficient bandwidth. Implementing this prioritization can significantly impact user experience, especially in environments with varied traffic types. Well-defined QoS policies prevent less critical traffic from overwhelming essential services, maintaining stable performance even under heavy load. Regular review and refinement of these policies are essential to adapt to your evolving workload.

Remember, network performance optimization is an ongoing process, not a one-time fix. Regular monitoring, proactive adjustments, and a keen eye for potential bottlenecks will ensure your virtualized environment continues to perform at its best, maximizing return on investment and maintaining high availability. By carefully planning, implementing, and consistently monitoring your virtual network, you can effectively avoid performance issues and ensure that your virtualized infrastructure runs smoothly and efficiently.

3.5 Troubleshooting Network Issues

Identifying the source of network problems in a virtualized environment can feel like navigating a maze. Start by checking the basics: are the virtual machines themselves powered on? Verify their network adapters are configured correctly, and that they have the appropriate IP addresses, subnet masks, and default gateways. A simple ping test to the default gateway or a known host on the same subnet can quickly pinpoint whether the issue lies within the VM itself or further down the network chain. Don't overlook the obvious—a simple cable disconnect can cause hours of frustration. Next, examine the virtual switch configuration. Are the virtual machines connected to the correct virtual switch? Is the virtual switch properly configured and linked to the physical network adapter? Incorrect settings here are common culprits for connectivity problems. Investigate the virtual switch's port groups; ensure that VMs are assigned to the correct port groups and that these groups have the necessary security policies and bandwidth allocations. Consider using a network monitoring tool to visually inspect the virtual switch's traffic flow; this can highlight bottlenecks or unusual activity.

If the problem persists within the virtual switch, expand your investigation to the hypervisor's network settings. Is the hypervisor itself correctly configured to access the physical network? Review the hypervisor's networking stack for errors or inconsistencies; incorrect configuration here can severely impact network performance and connectivity for all virtual machines. Ensure the hypervisor's network drivers are updated to the latest versions and check for any conflicts with other network interfaces or services. Restarting the hypervisor, while disruptive, can sometimes resolve mysterious network issues.

Beyond the hypervisor, the physical network infrastructure itself requires scrutiny. Check for physical cable issues, faulty network devices like switches and routers, and any network outages that might be affecting the physical network's connectivity. Employing a network analyzer to pinpoint congestion, packet loss or latency will provide invaluable data in resolving network issues within the physical infrastructure. Utilize tools like Wireshark or tcpdump to capture and analyze network traffic, allowing for detailed examination of individual packets to identify potential problems.

Now, consider the implications of network virtualization technologies like VLANs and VXLAN. If you're using these technologies, ensure that your VLANs are correctly configured and that the virtual machines are assigned to the appropriate VLANs. Misconfigurations in VLAN settings can lead to network segmentation issues, preventing VMs from communicating effectively. VXLAN, while offering scalability, also presents complexities. Make sure VXLAN tunnels are correctly established and that the underlying network infrastructure supports the VXLAN encapsulation and decapsulation processes.

Finally, let's not forget security. Firewalls and other security measures, both at the hypervisor and the physical network level, might be inadvertently blocking network traffic. Review your firewall rules to ensure that necessary ports and protocols are allowed through. Examine any intrusion detection or prevention systems (IDS/IPS) for any alerts or events that might indicate network intrusions or anomalies. An overlooked security rule, particularly within the hypervisor's firewall, could be subtly blocking connections while showing no obvious errors. Consistent monitoring and logging of security events are crucial to quickly respond to any network security incidents.

Remember, troubleshooting is iterative. Start with the simplest checks, progressively moving towards more complex investigations. Document your findings and the steps taken at every stage; this aids both in problem resolution and in preventing similar issues in the future. A comprehensive log can prove invaluable when dealing with complex or intermittent network problems. Thorough documentation of network configurations and troubleshooting steps is indispensable for efficient problem-solving in a virtualized environment. A methodical approach, along with the right tools and a bit of patience, can significantly improve your ability to resolve network issues in your virtualized infrastructure effectively and efficiently.

4 Storage Virtualization

Storage virtualization dramatically simplifies data management within a virtualized environment. By abstracting physical storage resources, it presents a unified, logical view to virtual machines, regardless of the underlying physical hardware. This allows for easier management, scalability, and improved resource utilization. Imagine managing terabytes of data spread across multiple physical storage devices - a daunting task. Storage virtualization transforms this complexity into a streamlined, manageable system.

Consider a scenario where a virtual machine requires additional storage. With traditional storage, this would involve physically adding a hard drive, a process that's time-consuming and potentially disruptive. With storage virtualization, however, you simply allocate more storage space from the virtual pool, a seamless process that happens within minutes. The underlying physical storage is automatically managed by the virtualization layer, ensuring optimal performance and minimal downtime. This is a significant advantage for businesses that require constant data accessibility and rapid scalability.

The benefits extend beyond mere convenience. Storage virtualization offers significant cost savings by optimizing storage utilization. Traditional storage often leads to wasted space due to under-utilization of individual drives. Storage virtualization pools resources, eliminating this inefficiency and allowing for dynamic allocation based on real-time needs. This ensures that storage resources are used effectively, minimizing capital expenditure and maximizing ROI. Moreover, consolidating storage onto fewer physical devices reduces power consumption and cooling requirements, contributing further to cost savings.

Implementing storage virtualization involves several key technologies. Storage area networks (SANs) provide a high-performance, block-level storage solution ideal for virtualized environments. SANs create a dedicated network for storage traffic, improving performance and scalability. Network-attached storage (NAS) offers a simpler, file-level approach, well-suited for applications that require shared access to files. The choice between SAN and NAS depends largely on specific needs and the nature of the applications running on the virtual machines. Thorough planning and understanding of your specific environment are crucial in making an informed decision.

The selection of the appropriate storage virtualization platform is also crucial. Different vendors offer various solutions with varying features and capabilities. Some platforms offer advanced features such as data deduplication and thin provisioning, which further enhance storage efficiency. Thin provisioning allows for virtual allocation of storage, only consuming physical space as data is written, maximizing space utilization. Data deduplication eliminates redundant data, reducing storage requirements and improving backup performance. These advanced features can significantly reduce the total cost of ownership over the long term. However, careful consideration of the platform's compatibility with your existing infrastructure is essential to prevent integration issues.

Effective management of a storage virtualized environment requires diligent monitoring and proactive maintenance. Regularly reviewing storage utilization, performance metrics, and capacity planning is crucial. Identifying and addressing potential bottlenecks before they impact operations is key to ensuring high availability and optimal performance. This involves establishing clear performance baselines and implementing alerts to notify administrators of potential issues. Tools built into the virtualization software, along with third-party monitoring solutions, offer comprehensive insights into the storage environment. Using this data wisely is key to optimizing performance and minimizing the risk of downtime.

Furthermore, disaster recovery planning is paramount in a virtualized environment. Regularly backing up virtual machines and their associated storage is crucial. Strategies include replicating storage to secondary locations, utilizing cloud-based backup solutions, or employing disaster recovery as a service (DRaaS). The choice depends on several factors such as budget, recovery time objectives (RTOs), and recovery point objectives (RPOs). Implementing robust disaster recovery plans is paramount to ensuring business continuity in the face of unexpected events. This minimizes potential financial losses and ensures the business continues to operate smoothly despite interruptions.

Finally, security within a storage virtualized environment should never be overlooked. Implementing robust security measures, such as access controls, encryption, and regular security audits, is crucial to protecting sensitive data. Appropriate security policies and procedures should be established and enforced to ensure data integrity and confidentiality. Regular vulnerability scans and penetration testing can help to identify and address potential security weaknesses. Keeping software up-to-date with security patches is critical, and it should be performed routinely. A layered approach to security is the most effective way to safeguard the storage environment and the data it contains. Remember, robust security practices are not just a best practice; they're a critical necessity.

4.1 SAN and NAS Virtualization

Virtualizing storage, specifically using Storage Area Networks (SANs) and Network Attached Storage (NAS), significantly boosts efficiency and scalability in virtualized environments. SANs, with their high-performance block-level storage, are ideal for demanding applications needing rapid data access. They offer centralized management and simplified data sharing across multiple VMs, resulting in improved resource utilization. Consider the impact on your workflow: imagine the time saved by not managing individual storage devices for each server. This centralized approach significantly reduces administrative overhead. NAS, conversely, provides file-level storage, often preferred for simpler applications or shared file access. Its ease of implementation and relatively lower cost make it an attractive option, particularly for smaller deployments or those with less demanding storage needs. The choice between SAN and NAS virtualization hinges upon specific application requirements and budgetary constraints. Think carefully about the I/O demands of your applications and your overall storage capacity needs. This decision isn't about picking a superior option; it's about selecting the most appropriate technology for your unique circumstances.

Understanding the underlying technologies is crucial. SANs usually employ Fibre Channel or iSCSI protocols, providing high bandwidth and low latency for optimized performance. These protocols, designed for speed and efficiency, are best suited for database applications or virtual desktops demanding consistent, fast response times. Think of it like this: for a high-speed racecar needing rapid acceleration, Fibre Channel would be the preferred fuel. Conversely, NAS typically uses network protocols like NFS or CIFS, readily accessible through standard network interfaces. The simplicity inherent in NAS's implementation is often its most significant advantage. You can easily integrate it into your existing network infrastructure without needing specialized hardware or expertise. Consider the impact on your IT team. A simplified implementation can mean less training, more efficiency, and more time to focus on other strategic tasks.

Effective virtualization of either SAN or NAS requires careful planning and implementation. You'll need a robust virtualization platform capable of handling the chosen storage type. Factors such as storage capacity, performance requirements, and scalability should be diligently assessed to prevent future bottlenecks. Proper planning avoids costly upgrades and potential downtime down the road. Think of this phase as laying the foundation of a building: A well-planned foundation ensures stability and longevity. Neglecting this leads to cracks and instability later. Begin by defining your storage needs - not just current, but future anticipated needs. Accurate forecasting is crucial to preventing unexpected issues. Regular performance monitoring will be critical to detecting potential problems and optimizing resource allocation.

Security considerations are paramount. Data security must be integrated into every step of the process. Implement access controls, encryption, and regular backups to protect sensitive information stored within the virtualized environment. Data breaches can be incredibly costly, both financially and reputationally. Remember the importance of the "defense in depth" strategy, adding multiple layers of security. Think of it like fortress walls: multiple layers provide more effective protection than a single wall. Consider factors like firewalls, intrusion detection systems, and regular security audits to ensure your virtualized storage is well-protected. Security should not be an afterthought but an integral part of your virtualization strategy from the initial planning stage.

Finally, remember that the choice between SAN and NAS isn't absolute. Hybrid approaches are possible and often beneficial. This combined approach can provide the best balance of performance, cost-effectiveness, and scalability. You might use SAN for critical applications demanding high performance and NAS for less demanding tasks like file sharing. Flexibility is key. As your needs evolve, you can adjust your storage solution to accommodate those changes. This iterative approach ensures that your virtualized storage remains efficient and adaptable to your ever-changing needs. Remember, technology changes rapidly, and your storage strategy should evolve to keep pace. The ability to adapt will be crucial to the long-term success of your virtualized infrastructure.

4.2 Storage Area Networks

Storage Area Networks (SANs) offer a centralized, high-performance approach to data storage, crucial for supporting the demands of virtualized environments. They provide a dedicated, high-speed network specifically designed for block-level storage access, unlike network-attached storage (NAS) which operates at the file level. This distinction is key; SANs deliver faster I/O operations, vital for resource-intensive virtual machines.

Imagine a scenario with multiple virtual servers demanding rapid access to large databases. A SAN seamlessly delivers that speed, minimizing latency and maximizing the performance of your entire virtual infrastructure. This efficiency directly translates to improved application responsiveness and a better user experience. The scalability of SANs is another compelling advantage; you can easily expand storage capacity as your needs grow, adding new storage arrays without disrupting existing operations. This adaptability is particularly beneficial in rapidly evolving virtual environments.

Moreover, SANs enhance data availability and disaster recovery capabilities. Through features such as RAID configurations and data replication, SANs provide redundancy and protection against data loss. In the event of a hardware failure, the SAN's built-in mechanisms ensure continuous operation, minimizing downtime and protecting valuable business data. This resilience is paramount in maintaining the high availability that's essential in a virtualized setting. Consider the cost savings resulting from reduced downtime; SANs contribute significantly to minimizing operational expenses.

Different SAN architectures exist, each with its strengths and weaknesses. Fibre Channel SANs, for example, have historically been the dominant technology, offering high bandwidth and low latency. However, iSCSI SANs, leveraging existing Ethernet infrastructure, provide a more cost-effective alternative. Understanding these architectural differences is critical for making informed decisions when designing your storage solution. The choice will depend on your specific requirements, budget, and existing network infrastructure. Careful consideration of these factors ensures a well-integrated and efficient storage solution.

SAN management is also crucial for optimal performance and reliability. Proper configuration, monitoring, and maintenance of your SAN are essential. Effective monitoring tools help identify potential bottlenecks and address issues proactively. Regular backups and disaster recovery planning are indispensable components of a robust SAN strategy. Proactive maintenance helps prevent costly downtime and data loss, ensuring your virtual environment's continued smooth operation. Don't overlook the importance of security; access controls and encryption are vital in safeguarding your valuable data.

Implementing a SAN is a significant undertaking. Careful planning is crucial to ensure a successful integration. This includes assessing your current storage needs, projecting future growth, and selecting appropriate hardware and software. Thorough testing before deploying the SAN into a production environment is strongly recommended. This mitigates potential issues and ensures seamless integration with your existing virtual infrastructure. The transition to a SAN architecture often requires expert assistance; consulting with experienced IT professionals can streamline the process and avoid costly mistakes.

Beyond the technical aspects, the financial implications of SAN adoption must be carefully considered. While the initial investment can be substantial, the long-term benefits often outweigh the costs. Improved performance, increased data availability, and reduced downtime can significantly impact your operational efficiency and profitability. A detailed cost-benefit analysis, factoring in the total cost of ownership, is an essential step in the decision-making process. Remember, choosing the right SAN solution is a strategic investment that directly influences the overall success of your virtualized environment. Think of the long-term ROI; it's an investment that pays dividends in efficiency and reliability.

Finally, remember that your choice of SAN should directly reflect the needs of your virtualization strategy. A well-planned SAN is not just a storage solution, it's a fundamental pillar supporting the stability and performance of your entire virtual infrastructure. A poorly planned or inadequately managed SAN can become a significant bottleneck, undermining the very benefits that virtualization is intended to deliver. Therefore, selecting a SAN that aligns with your overall virtualization goals is a critical decision that needs careful consideration and planning. Invest wisely, and your virtual environment will thrive.

5 High Availability Strategies

High availability is paramount in any virtualized environment. Downtime translates directly to lost productivity and revenue, making robust strategies crucial for success. Building a system designed for resilience requires careful planning and implementation, starting with a deep understanding of your application's needs and potential failure points.

Redundancy is the cornerstone of high availability. This involves creating backups for critical components - servers, storage, and network infrastructure. Consider implementing RAID configurations for storage, providing data protection against drive failures. Multiple network connections can prevent outages caused by network failures. These redundant systems act as safeguards, ensuring minimal disruption should a primary component fail. The cost of implementing redundancy needs to be weighed against the potential cost of downtime; a thorough risk assessment helps determine the appropriate level of redundancy.

Clustering goes beyond simple redundancy; it actively distributes workload across multiple servers. In a cluster, multiple virtual machines can run simultaneously on different physical servers. If one server fails, the virtual machines automatically fail over to another, ensuring continuous operation. The implementation of clustering involves choosing the right software and hardware, configuring the cluster for optimal performance, and rigorously testing the failover process. This ensures a seamless transition with minimal interruption to users. Understanding the different clustering technologies - like VMware HA or Microsoft Failover Clustering - is essential for making informed decisions aligned with your specific needs and budget.

Disaster recovery planning extends high availability beyond simple component failures. It addresses larger-scale events, such as natural disasters or widespread power outages. A robust plan should outline procedures for backing up data to offsite locations, relocating virtual machines to a secondary data center, and restoring services in the event of a catastrophic failure. Regular testing of your disaster recovery plan is essential to ensure its effectiveness and identify any potential weaknesses or gaps. The plan should encompass all aspects of the virtualization infrastructure including network and storage, specifying roles, responsibilities, and communication protocols. Remember, a well-rehearsed plan is critical for swift and effective recovery.

Proper monitoring is vital for maintaining high availability. Real-time monitoring of critical system parameters, like CPU utilization, memory usage, and network traffic, helps identify potential issues before they cause significant problems. Setting up appropriate alerts ensures timely intervention. The data gathered through monitoring provides valuable insights for optimization, enabling proactive adjustments to resource allocation and system configuration, and preventing potential bottlenecks that could jeopardize availability. Effective monitoring solutions need to be integrated into your existing infrastructure, and staff needs to be trained to interpret the data and take appropriate action.

Implementing high availability isn't a one-size-fits-all solution. Best practices necessitate understanding your specific needs and priorities. A cost-benefit analysis is key, balancing the investment in high availability technologies with the potential cost of downtime. Prioritize critical applications and services. Focus on those applications that have the highest impact on your business operations, directing resources and effort where they will have the greatest effect. Thoroughly document your high availability strategy, detailing procedures, contact information, and escalation paths. This documentation ensures a consistent response to issues and facilitates knowledge sharing amongst the team. Regular reviews and updates ensure that your plan remains aligned with evolving needs and vulnerabilities, reflecting the dynamic nature of technology. Continuous learning is vital, staying informed about best practices and emerging technologies will ensure your high availability strategy remains robust and efficient.

5.1 Clustering and Failover

High availability is paramount in today's always-on world. Downtime translates directly to lost revenue and compromised productivity, making robust strategies crucial. Clustering and failover mechanisms are your primary defenses against this. Imagine a scenario where one server in your virtualized environment fails; with proper clustering, other servers seamlessly take over, ensuring continuous operation. That's the power of high availability.

Consider a simple two-node cluster. Each node runs identical virtual machines, with one acting as the primary and the other as a standby. If the primary node fails, the standby instantly assumes control, minimizing disruption. This is the core of failover - the automatic transfer of operations to a backup system when the primary system fails. Monitoring plays a key role; sophisticated systems constantly check the primary node's health, triggering failover if any issues are detected.

Implementing clustering requires careful planning. First, you must choose the right clustering technology, considering your specific needs and the compatibility with your virtualization platform. Popular options include VMware HA, Microsoft Failover Clustering, and various open-source solutions. Each has its strengths and weaknesses, so selecting the best one depends entirely on your infrastructure and budget. Before implementation, thorough testing in a non-production environment is essential. This allows you to identify and resolve any potential issues before they affect your live systems.

Beyond the basic two-node configuration, more complex cluster setups are possible. These often involve multiple nodes working together, offering enhanced redundancy and scalability. For instance, a three-node cluster provides even greater resilience, as a single node failure leaves two still operational. This enhanced redundancy, coupled with sophisticated load balancing, allows the cluster to handle much larger workloads while maintaining consistent performance. Furthermore, the design should consider the specific applications running on the virtual machines. Some applications are more tolerant to downtime than others, influencing the design choices for redundancy levels and failover times.

Network considerations are equally important. High-speed, reliable networking is the backbone of any successful clustering strategy. Network latency can significantly impact failover time, meaning a slow network can lead to noticeable downtime even with a well-designed cluster. A robust network infrastructure with redundant pathways is therefore essential to ensure fast and reliable communication between cluster nodes. This involves redundant switches, routers, and cabling to prevent single points of failure from disrupting communication among the servers within the cluster.

Storage is another critical aspect. Shared storage, accessible to all nodes in the cluster, is usually the preferred method. This ensures that if a node fails, the other nodes can immediately access the necessary data and continue operations without interruption. However, it's not always necessary and different scenarios call for different approaches. Storage area networks (SANs) and network-attached storage (NAS) can be leveraged to provide the shared storage capability. The choice depends on several factors, including budget, performance requirements, and scalability.

Regular testing is vital. Simulating node failures allows you to verify that the failover mechanism works as expected, identify any potential bottlenecks, and ensure minimal downtime. This testing shouldn't be a one-time event, but an ongoing process incorporated into your regular maintenance schedule. Regular testing can also aid in refining the failover process, identifying areas for optimization, and ensuring overall system robustness. This proactive approach ensures your high availability strategy is always at peak efficiency. Remember, the goal is to minimize downtime to a point that's essentially imperceptible to end users, and that requires ongoing vigilance and meticulous testing.

Finally, remember that clustering and failover are just two pieces of a broader high-availability puzzle. Other important elements include regular backups, disaster recovery planning, and robust monitoring systems. While clustering and failover provide immediate protection against node failures, these additional measures add another layer of protection against wider disruptions, ranging from hardware failures to natural disasters. A comprehensive high-availability strategy integrates all these elements to provide a resilient and reliable virtualized environment, prepared for nearly any eventuality. The integration of these different strategies is crucial; a single point of failure in any part of the system can jeopardize the entire high availability strategy. Careful attention to each detail ensures a robust, reliable, and cost-effective solution.

5.2 Redundancy and Disaster Recovery

Redundancy and disaster recovery are paramount for maintaining business continuity in a virtualized environment. A well-defined strategy minimizes downtime and data loss, protecting your valuable assets. This isn't merely about having backups; it's about building a resilient system that can withstand various failures.

Implementing redundancy involves creating duplicate components or systems. Think of it as having a spare tire for your car—you hope you never need it, but it's invaluable when you do. In virtualization, this might involve having duplicate virtual machines, network connections, or storage arrays. This duplication ensures that if one component fails, another immediately takes over, seamlessly ensuring service continuity. Consider using geographically dispersed data centers, so that even a regional disaster won't cripple your operations.

High availability, closely linked to redundancy, goes beyond simple duplication. It's about designing your system to handle failures gracefully and automatically. Clustering technologies, for example, allow multiple servers to work together, with one taking over if another fails. This approach minimizes the impact of individual server failures on overall system performance. Regular testing of your failover mechanisms is essential; it's not enough to have a backup; you need to know it works.

Disaster recovery plans, often overlooked, are crucial for larger-scale incidents. These plans outline procedures to restore your entire virtualized infrastructure in the event of a catastrophic event—a fire, flood, or even a widespread cyberattack. Your plan must account for data backup and restoration procedures, system recovery timelines, and communication protocols during an emergency. Don't just create a document; actively test the plan regularly to ensure its efficacy and identify any weaknesses. The goal isn't just to get back online; it's to get back online quickly with minimal data loss.

Choosing the right replication strategy depends heavily on your specific requirements and tolerance for downtime. Synchronous replication, where data is written to multiple locations simultaneously, offers the highest level of protection but comes at the cost of reduced performance. Asynchronous replication, where data is written to secondary locations periodically, offers better performance but slightly increases the risk of data loss. Evaluate the tradeoffs carefully based on your Recovery Time Objective (RTO) and Recovery Point Objective (RPO).

Beyond the technical aspects, consider the human element. Your disaster recovery plan must include clear roles and responsibilities for your team. Who's responsible for what? What communication channels will you use during an emergency? How will you keep your stakeholders informed? Regular training and drills are essential to ensure that everyone knows their role and can execute the plan effectively. This isn't just about technology; it's about people working together. Documenting these procedures is also crucial for maintainability and continuity if personnel changes occur within your team. Remember, redundancy and disaster recovery aren't one-time setups; they're ongoing processes. Regularly review and update your plans, as your infrastructure evolves and your business needs change. Technology advances, vulnerabilities emerge, and your RTO and RPO may shift. Staying proactive is key to maintaining a resilient virtualized environment. The cost of inaction far outweighs the investment in robust redundancy and disaster recovery strategies. It's not a matter of if disaster will strike, but when. Be prepared.

5.3 High Availability Best Practices

Implementing robust high availability (HA) isn't just about deploying redundant components; it's about a holistic approach encompassing proactive planning, meticulous execution, and ongoing monitoring. Start with a comprehensive risk assessment, identifying critical applications and their dependencies. This foundational step dictates your HA strategy, guiding choices regarding redundancy levels and recovery time objectives (RTOs) and recovery point objectives (RPOs). Failing to prioritize correctly will leave your most crucial systems vulnerable, undermining the entire effort.

Next, consider the various HA strategies available. Clustering, for instance, provides a powerful way to ensure uninterrupted service through automatic failover. However, choosing the right clustering technology—be it a shared-nothing, shared-disk, or other architecture—depends heavily on factors such as application requirements, budget constraints, and the existing infrastructure. Careful consideration of these factors is vital to avoid costly mistakes and ensure a seamless transition. Poorly chosen technology can lead to complexities and compatibility issues, increasing downtime rather than preventing it.

Redundancy is another cornerstone of HA. This means having backups for every critical component: servers, network devices, storage systems. This isn't just about having spare hardware; it's about having a well-defined procedure for switching over to those backups in case of failure. This includes rigorous testing of the failover process, not just in a simulated environment, but also during off-peak periods in the live environment. Neglecting testing will expose your system to unforeseen complications during a real emergency.

Disaster recovery (DR) planning goes beyond simple redundancy. It's about defining a comprehensive strategy for recovering your entire virtual infrastructure in the event of a catastrophic failure, such as a natural disaster or a significant data center outage. This involves establishing a secondary site, either geographically dispersed or a cloud-based solution. The DR plan must include detailed procedures for data backup, replication, and recovery—with regular testing and updates to reflect changes in your infrastructure and business needs. A poorly planned DR strategy will severely limit your capacity to recover from unexpected events, leading to significant financial and reputational damage.

Beyond the technological aspects, strong operational practices are paramount. Regular monitoring of your virtual infrastructure is crucial. Employ proactive monitoring tools to identify potential problems before they escalate into outages. These tools should provide real-time alerts, enabling swift intervention and preventing minor issues from snowballing. Implementing a robust alert system is not sufficient; you need a dedicated team trained to respond efficiently to these alerts. An unprepared or unresponsive team renders even the best technology ineffective.

Furthermore, comprehensive documentation is essential for effective HA. This includes detailed descriptions of your infrastructure, HA strategies, and recovery procedures. This documentation should be easily accessible to all relevant personnel. Clear, concise documentation ensures that everyone involved understands their roles and responsibilities in the event of a failure. Insufficient or poorly maintained documentation will lead to confusion and delay during a critical recovery process.

Finally, regular training and drills are crucial. Your team needs to be thoroughly familiar with all aspects of your HA strategy, including the use of monitoring tools, failover procedures, and recovery processes. Regular drills, conducted under simulated failure scenarios, help ensure everyone is prepared and coordinated, minimizing response time and preventing human error. Insufficient training can negate the benefits of a perfectly implemented HA solution, leaving your system vulnerable during an actual crisis. A well-rehearsed team is your most valuable asset in maintaining high availability. The integration of all these elements—proactive planning, appropriate technology choices, robust redundancy, comprehensive disaster recovery, continuous monitoring, thorough documentation, and regular training—creates a resilient and highly available virtual environment. Remember, high availability isn't a one-time project; it's an ongoing process requiring continuous attention and improvement.

6 Cost Optimization Techniques

Efficient resource utilization is paramount. Consolidating multiple virtual machines (VMs) onto fewer physical servers significantly reduces hardware costs, including power consumption and server maintenance. Imagine, instead of ten underutilized servers, you manage just three, each running a carefully planned mix of VMs. This not only saves money directly but also simplifies management and reduces the physical space needed. Careful planning and monitoring are crucial for successful consolidation, preventing performance bottlenecks and ensuring application stability. Strategic VM placement, considering resource demands and dependencies, is key to maximizing this approach's benefits.

Right-sizing your VMs is equally important. Over-provisioning, providing VMs with more resources than needed, leads to wasted resources and increased costs. Conversely, under-provisioning might result in performance issues, impacting productivity and possibly leading to costly application downtime. Start by thoroughly analyzing your application's resource requirements. Utilize performance monitoring tools to track CPU, memory, and I/O usage. Adjust VM resource allocations based on these data points, ensuring they meet demands without excessive over-provisioning. Regularly review VM performance metrics and make adjustments as application requirements evolve, striving for that sweet spot of optimal performance and cost-effectiveness.

Energy efficiency plays a vital role in long-term cost savings. Power consumption is a considerable expense in large-scale virtualized environments. Employing energy-efficient hardware such as servers with high power utilization efficiency (PUE) ratings is a direct way to reduce electricity costs. Virtual machine consolidation, discussed earlier, directly contributes to energy savings as well. Beyond hardware, implement power management strategies, such as scheduling VM power-on and power-off times to align with usage patterns, minimizing idle periods, further reducing the energy burden. These seemingly small actions add up considerably.

Software licensing costs can easily escalate. Optimize your licensing strategy by accurately assessing your needs and avoiding over-purchasing. Implement thorough license tracking systems to monitor utilization rates and ensure compliance. Explore options like using licensing models that accommodate fluctuating needs, such as pay-as-you-go licenses or subscriptions. Consider open-source alternatives for certain software components to further decrease expenses. Proactive license management is crucial, preventing unexpected expenses and penalties. Regular auditing and reporting helps maintain control over these often complex and potentially expensive items. Automation is a powerful tool for cost optimization. Automating routine tasks, like VM provisioning, patching, and backups, reduces manual effort, freeing IT staff for more strategic initiatives. This increased efficiency translates to significant cost savings. Automation also reduces human error, minimizing the chances of costly mistakes. Invest in automation tools, and implement procedures that simplify these often laborious tasks. The increased efficiency will help keep your costs down and will ultimately improve your team's productivity. The initial investment in automation pays off handsomely in the long run through reduced operational expenses and improved resource allocation.

6.1 Resource Consolidation

Efficient resource utilization is paramount in virtualization. Think of your virtual infrastructure as a city—optimizing its space and resources directly impacts its overall functionality and cost-effectiveness. Consolidating resources involves strategically grouping VMs onto fewer physical servers, maximizing hardware utilization, and minimizing energy consumption. This isn't simply about stuffing more VMs onto a single server; rather, it's about intelligent placement, considering factors like VM resource demands, application criticality, and potential performance bottlenecks. Effective consolidation requires careful planning and analysis, but the rewards -

reduced hardware costs, lower energy bills, and simplified management - are substantial.

Careful analysis of your current VM environment is crucial before starting consolidation. This isn't a haphazard process. You need to identify underutilized servers, those with significant headroom in terms of CPU, memory, and storage. Use monitoring tools to gain a comprehensive picture of your VMs' resource consumption patterns. Look for consistent underperformance across several machines. These are prime candidates for consolidation onto fewer, more powerful hosts. Remember to consider future growth when making these assessments; a proactive approach is essential to avoiding resource constraints later. Consider deploying a comprehensive monitoring system that automatically alerts you to approaching resource thresholds, allowing you to preemptively adjust your resource allocation before performance issues arise. This forward thinking prevents unexpected downtime.

Once you've identified target VMs for consolidation, you'll need to carefully plan the migration process. This involves creating a detailed migration plan, outlining the order of VM movements, the time windows for migrations, and the necessary downtime. Test the migration process thoroughly in a non-production environment to identify and address any potential issues. Remember, meticulous planning is key; an unplanned or rushed migration could easily lead to outages or data loss, negating the benefits of consolidation. For instance, carefully planned downtime can significantly mitigate the risk of user disruption. Prioritize essential VMs and schedule migrations during off-peak hours.

The next phase involves the actual migration of VMs. Leverage your virtualization platform's features for live migration or vMotion capabilities if available. This allows for seamless VM movement without any noticeable downtime. If live migration isn't feasible, consider using tools to create snapshots or backups before the migration to minimize the risk of data loss. Regardless of the approach, meticulous documentation and careful execution are crucial. Should you encounter errors, it is critical to roll back immediately and review the steps to prevent future problems. The goal is a smooth transition. Always back up critical data before starting the consolidation procedure.

Post-consolidation, monitoring is critical to ensure that the consolidated environment is stable and performing as expected. Continue to monitor resource utilization, paying close attention to CPU, memory, and storage usage on the consolidated servers. Adjust resource allocation as needed to maintain optimal performance. You might find that some VMs require more resources than initially anticipated, necessitating adjustments to resource pools or even the addition of extra physical servers. Proactive monitoring is far better than reactive problem-solving; continuously reviewing your resource utilization allows for immediate adjustments. Regular monitoring should become a standard practice.

Finally, consider automation to streamline the resource consolidation process. Many virtualization platforms offer automation tools that can automate aspects of VM migration, resource allocation, and monitoring. By automating routine tasks, you free up valuable time and resources, reducing operational expenses. Automation reduces human error and improves efficiency. For example, implementing automated alerts for resource thresholds can significantly reduce the risk of performance bottlenecks. Consider integrating automation tools into your overall infrastructure management strategy for greater efficiency. Remember, efficiency improvements are not merely about cost reduction but also about enhancing your ability to adapt to change and growth within your infrastructure.

6.2 Right-Sizing VMs

Right-sizing virtual machines (VMs) is crucial for optimizing resource utilization and minimizing costs. Over-provisioning VMs leads to wasted resources and higher energy consumption, while under-provisioning can result in performance bottlenecks and application failures. Finding the sweet spot requires careful analysis and proactive management. It's a balancing act, a delicate dance between sufficient resources and unnecessary expenditure.

Begin by thoroughly assessing the resource requirements of each application running on your VMs. This involves analyzing CPU usage, memory consumption, disk I/O, and network bandwidth over a period of time, ideally during peak usage. Tools like performance monitors built into your hypervisor or third-party monitoring software can provide valuable insights. Don't rely on guesswork; solid data is your foundation. This detailed analysis provides a clear picture of the actual needs, separating the hype from the reality, allowing for more informed decisions.

Once you have a clear understanding of resource consumption patterns, you can start resizing your VMs. If a VM consistently shows low resource utilization across all metrics—well below the allocated capacity—reducing its size is a straightforward optimization. Conversely, a VM consistently hitting resource limits, exhibiting high CPU usage, memory pressure, or slow disk I/O, necessitates an increase in allocated resources. This isn't a one-time process; continuous monitoring is essential. Remember to consider future growth; some applications might experience fluctuating workloads, demanding extra capacity during peak times. Therefore, always leave room for growth and potential spikes.

The process of right-sizing isn't just about adjusting CPU and memory; it also extends to storage capacity and network bandwidth. A VM might need more disk space for data storage or a faster network connection for high-throughput applications. For instance, a database server will require significant storage, potentially necessitating a larger disk, while a web server might benefit from increased network bandwidth for handling concurrent requests. Regularly review these aspects alongside CPU and RAM to avoid unexpected performance drops.

Remember that the goal is not to squeeze every ounce of performance out of the lowest possible VM configuration. A slightly over-provisioned VM that operates smoothly and reliably can actually be more cost-effective in the long run than a constantly strained, under-provisioned one. The latter often leads to performance issues, requiring more troubleshooting and potential downtime, which may end up exceeding the cost savings from minimizing resources. Maintaining a buffer prevents disruptions and unnecessary administrative overhead.

Implementing a systematic approach is key. Don't try to right-size all your VMs at once. Start with a small pilot program, focusing on a subset of VMs representing different workload profiles. This allows you to test your procedures and refine your approach before scaling it across your entire virtual infrastructure. Document your findings and share your experiences to inform future decisions, thus establishing best practices and a consistent methodology. This gradual approach minimizes risks and allows for adaptation based on observed results.

Automate as much of the right-sizing process as possible. Many hypervisors offer tools and APIs to automate VM resizing. Scripting tools can be used to automate the monitoring, analysis, and resizing processes, saving you considerable time and effort. Automation reduces manual intervention, minimizes the margin for error, and increases the efficiency of the whole operation. Such automation is especially useful when dealing with a large number of VMs.

Finally, consider the cost implications of both over-provisioning and under-provisioning. Over-provisioning results in wasted resources and increased energy bills. Under-provisioning, on the other hand, can lead to performance problems, impacting application availability and potentially costing more in the long run due to downtime, decreased productivity, and potential business losses. Continuously monitoring and adjusting VM resource allocation is a critical element of cost-effective virtualization. A balanced approach, guided by data and informed by your specific needs, delivers the optimal balance between performance and cost. Through diligent monitoring and appropriate adjustments, you can achieve significant cost savings without compromising the stability and performance of your virtualized environment.

6.3 Energy Efficiency

Power consumption is a significant operational cost in any data center, and virtualized environments are no exception. Minimizing energy use directly translates to lower electricity bills and a reduced carbon footprint. Focusing on energy efficiency isn't just environmentally responsible; it's financially prudent. Smart energy management can significantly enhance the overall return on investment of your virtualization project.

Effective cooling strategies are crucial. Overheating is a leading cause of server failures and increased energy consumption. Invest in high-quality cooling systems, ensuring proper airflow and maintaining optimal temperatures within your server room. Regular maintenance of these systems is also paramount—preventative measures are far less costly than emergency repairs. Consider using air-cooled servers where appropriate, avoiding liquid cooling unless absolutely necessary for high-density deployments, which can complicate maintenance and increase initial infrastructure costs. Remember to regularly clean dust and debris from cooling components, which can significantly impact their effectiveness.

Virtual machine density optimization plays a key role. Consolidating multiple VMs onto fewer physical servers is a powerful way to reduce your energy footprint. This can be achieved through careful planning and VM right-sizing. Avoid over-provisioning resources, as it leads to wasted energy and capacity. Right-sizing VMs to meet their actual needs reduces unnecessary resource allocation; a server that's only half-utilized consumes far more energy per unit of processing than a fully loaded server. Regularly review VM resource usage, identifying opportunities for consolidation or downsizing. Tools and monitoring systems can automatically flag VMs for optimization based on pre-defined thresholds. The aim is to maximize the utilization of each physical server, reducing the overall number required and, consequently, power consumption.

Power management features offered by your virtualization software and server hardware should be fully utilized. These features often include power-saving modes that reduce power consumption during periods of low activity. Scheduling power-down during off-peak hours or for less frequently used VMs can yield considerable savings. Similarly, configuring servers to enter low-power states when idle can dramatically decrease energy consumption. Consider employing features like dynamic power capping, which automatically adjusts the power supply based on real-time demand. This intelligent approach ensures that power usage is proportional to the actual workload, minimizing waste. Remember to regularly review and adjust these settings based on usage patterns and performance requirements.

Server hardware selection is critical. Choose energy-efficient servers with features such as power-saving processors and low-power components. Look for servers certified with energy efficiency standards, such as the Energy Star rating. These certifications provide an objective measure of a server's power consumption, enabling informed decisions during procurement. While initial costs might be slightly higher for energy-efficient hardware, the long-term savings in electricity costs easily offset these initial investments. Furthermore, consider the lifespan and future maintenance of the hardware; a robust, energy-efficient system will often translate into lower maintenance costs over its operational life.

Implementing a comprehensive energy management strategy necessitates ongoing monitoring and evaluation. Utilize energy monitoring tools to track power consumption in real-time, identifying trends and areas for improvement. This data is essential for making informed decisions about resource allocation and optimization strategies. Regular analysis of this data will reveal inefficiencies and highlight opportunities for further optimization. By proactively monitoring energy consumption, you can quickly identify and address issues before they become major problems, potentially averting significant energy waste and ensuring the ongoing efficiency of your data center. The continuous monitoring also facilitates a feedback loop, allowing you to adjust your strategies and processes, progressively refining your energy efficiency initiatives. The result is not just significant cost savings but also the creation of a more sustainable and environmentally responsible data center operation.

6.4 Licensing and Cost Management

Understanding licensing agreements is paramount. Ignoring the fine print can lead to unexpected expenses and legal complications, significantly impacting your overall cost efficiency. Always thoroughly review the terms and conditions, paying close attention to the permitted number of virtual machines, the types of hosts allowed, and any limitations on usage. Consider engaging legal counsel if you're uncertain about any aspect of a license agreement; this seemingly small investment can prevent far larger financial losses down the line. Proactive legal review offers peace of mind and avoids potential costly disputes.

Negotiation is a powerful tool, often overlooked. Don't be afraid to negotiate with vendors for better licensing terms, especially for large-scale deployments. Bulk discounts are common, and vendors are often willing to work with clients to find mutually beneficial agreements. Clearly outlining your needs and the scale of your implementation provides leverage in negotiations. Remember to document all agreed-upon terms and conditions meticulously, ensuring both parties have a clear understanding of the commitments made. Thorough documentation protects you from future disagreements or misunderstandings.

Effective license management is more than just purchasing software. It's about actively tracking your licenses, ensuring compliance, and identifying opportunities for optimization. Implement a robust inventory system, whether it's a spreadsheet or a dedicated software solution, to track your licenses, expiration dates, and usage. Regular audits of your license inventory allow you to identify underutilized licenses that can be repurposed or returned, reducing waste and saving significant costs. Regular auditing is crucial for cost control and compliance.

Choosing the right licensing model is critical. Different licensing models offer varying levels of flexibility and cost structures. Carefully evaluate your needs and choose the model that best aligns with your budget and operational requirements. Consider subscription-based models for their flexibility and predictable costs. Alternatively, perpetual licenses may be more cost-effective in the long run if your usage remains consistent. A thorough cost-benefit analysis for each model is vital in the long-term cost savings.

Software asset management (SAM) tools can significantly improve your licensing efficiency. These tools automate license tracking, reporting, and compliance checks, freeing up IT staff for more strategic tasks. SAM tools provide comprehensive insights into license utilization, helping identify areas for optimization and cost reduction. Investing in a quality SAM tool is a long-term investment that pays for itself by streamlining processes and lowering costs. They offer a high return on investment through better management of software assets.

Cloud-based virtualization offers both advantages and challenges regarding licensing. Cloud providers frequently offer different licensing models for their services. You might pay per-use, per-instance, or based on other metrics. Carefully review the pricing structures of different cloud providers and compare them to on-premises licensing costs before making a decision. A proper comparison of cloud versus on-premise licensing models is crucial for effective cost management. Understand how your cloud provider's licensing policies affect compliance and scalability.

Open-source alternatives should always be considered. Open-source virtualization solutions can offer significant cost savings compared to proprietary software, although they may require more technical expertise. Evaluate open-source options based on their capabilities, community support, and suitability for your specific use case. Before implementing, carefully research the level of technical support and community resources available to ensure a seamless transition and efficient troubleshooting. Open-source software can significantly reduce licensing costs in the long run.

Regular review and optimization are essential. Regularly review your licensing agreements and utilization patterns. Identify areas where you can consolidate licenses, negotiate better terms, or transition to more cost-effective solutions. Continuously monitoring and adapting your licensing strategy ensures that you maintain an efficient and cost-effective virtualized environment. Continuous monitoring and optimization is key to long-term cost reduction. The initial setup isn't the only time to optimize costs, it's ongoing.

6.5 Automation and Efficiency

Automating tasks within your virtualized environment is key to maximizing efficiency and minimizing operational costs. Think of it like this: Manually configuring dozens of virtual machines is time-consuming and prone to errors. Automation streamlines this process, ensuring consistency and reducing human intervention. Tools like Ansible, Puppet, or Chef can automate VM creation, configuration, and deployment, drastically reducing the time and effort required. Furthermore, scheduled tasks, handled via scripting or dedicated tools, can perform routine maintenance such as patching, backups, and monitoring, all without manual intervention. This proactive approach ensures consistent performance and prevents potential issues before they impact productivity.

Implementing automated monitoring is another significant step towards efficiency. Imagine constantly checking the performance of each VM—exhausting, right? Sophisticated monitoring tools, integrated with your automation framework, provide real-time insights into resource utilization, performance bottlenecks, and potential failures. This allows for proactive problem solving and resource allocation, optimizing your virtual infrastructure's performance. Alerts can be automatically triggered based on predefined thresholds, notifying administrators of issues needing attention. This ensures quick responses and minimizes downtime, leading to a more robust and efficient virtualized environment.

Efficient resource allocation is paramount. Automation helps you achieve this. Scripting and automated tools can analyze resource usage patterns and dynamically allocate resources to VMs based on current needs. This prevents resource wastage and ensures that critical applications always have the necessary resources, without over-provisioning. This dynamic allocation strategy contrasts with static configurations, leading to a more cost-effective use of resources like CPU, memory, and storage. Furthermore, automation helps with right-sizing VMs, ensuring each VM is appropriately sized for its workload. Over-provisioning wastes resources and increases costs. Conversely, under-provisioning can lead to performance bottlenecks and decreased efficiency. Automated tools and scripting can analyze VM performance data and automatically adjust VM resources as needed.

Beyond VM management, automation significantly improves network and storage virtualization processes. Automated network configuration scripts can rapidly provision and configure virtual networks, simplifying network administration and reducing the risk of human error. Similar automation capabilities extend to storage management: automated provisioning and scaling of storage resources ensure that applications always have the storage they need, simplifying storage administration and preventing bottlenecks. These automated processes aren't just about saving time; they also improve consistency and reduce the likelihood of misconfigurations. In large-scale environments, consistent configuration is crucial for maintainability and reliability.

Consider the implications of integrating automation with your backup and recovery strategy. Automated backup scheduling and testing dramatically reduce the risk of data loss. In the event of a disaster, automated recovery scripts can restore services quickly and efficiently, minimizing downtime and ensuring business continuity. This proactive and automated approach is superior to manual processes, which are more time-consuming, error-prone, and often lack the consistency necessary for reliable disaster recovery. Automated testing of your backups further increases confidence in the reliability of your recovery strategy. Regular testing ensures that backups are valid and can be restored successfully.

Finally, remember that continuous improvement is essential. Regularly review your automation processes, identifying areas for improvement and optimization. As your virtual environment evolves, so should your automation framework. Analyze the performance of your automated scripts and processes, using collected data to refine and improve their effectiveness. Don't hesitate to adapt your approach based on evolving needs and technological advancements. The best automation strategies are flexible and adaptive, responding to the ever-changing landscape of virtualization technology. Regular review and fine-tuning are essential for maintaining an efficient and robust system.

7 Security in Virtualized Environments

Securing your virtualized environment isn't just about protecting individual virtual machines; it's about creating a robust, layered defense across your entire infrastructure. Think of it like building a castle - you need strong walls, vigilant guards, and a well-planned escape route in case of a breach. Neglecting any aspect weakens the whole system. Ignoring security at the outset can lead to significant vulnerabilities, making your data and your business vulnerable to attacks, and that's a costly mistake, far exceeding any savings gained through virtualization. Remember, the benefits of virtualization are negated if your system is compromised.

Implementing strong passwords and multi-factor authentication is fundamental. Don't underestimate the power of this simple, yet incredibly effective measure. It's the first line of defense, preventing unauthorized access to your VMs and the underlying infrastructure. Regular password changes, coupled with a robust password policy that enforces complexity, are crucial. Beyond passwords, consider implementing multi-factor authentication (MFA), adding an extra layer of security that requires more than just a password to gain access. This might involve using a mobile app, security key, or another verification method. It's simple to implement, yet immensely powerful in preventing unauthorized access.

Regular patching and updates are paramount. Operating systems, hypervisors, and applications all have vulnerabilities; neglecting updates exposes your system to known exploits. Develop a rigorous patching schedule, testing updates in a non-production environment first before deploying them to live systems. This phased approach minimizes disruption and allows for quick identification and resolution of any unforeseen issues. Automation tools can streamline this process, freeing up your IT team to focus on other crucial tasks. Remember, timely updates are your first line of defence against emerging threats.

Network segmentation is vital for isolating VMs and limiting the impact of a security breach. Instead of having all your VMs on a single network, create separate virtual networks based on function and sensitivity of data. This way, if one VM is compromised, the attacker's access is restricted to that specific network segment. Implementing robust firewalls between these segments further enhances security, creating controlled access points and minimizing the risk of lateral movement within the infrastructure. This architectural approach drastically reduces the impact of potential threats.

Regular security audits and penetration testing provide critical insights into the effectiveness of your security measures. Independent audits provide an objective assessment of your vulnerability profile, identifying weaknesses before attackers do. Penetration testing simulates real-world attacks, highlighting potential gaps in your security posture. The results of these exercises help prioritize remediation efforts, ensuring your defenses remain strong against the ever-evolving threat landscape. It's an investment that pays for itself in the long run, significantly reducing the risk of a successful attack. Virtual machine sprawl can create unnecessary complexity and increase your attack surface. Ensure you consistently consolidate and right-size your VMs, removing those that are unused or underutilized. This practice simplifies management, reduces the number of potential entry points for attackers, and optimizes resource consumption. Regular review and cleanup of your VM infrastructure helps prevent security issues arising from neglected or poorly managed resources. A well-managed virtual environment reduces risks, simplifies security administration, and lowers costs.

Finally, robust data protection strategies are crucial. This includes implementing data encryption both in transit and at rest, using strong encryption algorithms and regularly reviewing and rotating encryption keys. Regular backups are essential to recover from data breaches or system failures. Ensure your backup strategy is comprehensive, tested, and secure, including measures to prevent unauthorized access to backup data. This multi-layered approach ensures data protection even in the event of a major system failure or security incident. Remember, data protection goes beyond security; it's about business continuity.

By employing these strategies, you'll build a significantly more secure virtualized environment, protecting your valuable data and your business. Remember that security is an ongoing process, not a one-time fix. Continuously monitor your systems, stay updated on the latest threats, and adapt your security measures accordingly. The investment in robust security is far less costly than dealing with the aftermath of a breach. Proactive security is the key to a secure and thriving virtualized environment.

7.1 Securing Virtual Machines

Regular patching is paramount. Outdated software represents a gaping vulnerability, readily exploited by malicious actors. Prioritize patching your hypervisor, guest operating systems, and any applications running within the VMs. Implement a robust patching schedule, testing updates in a non-production environment before deploying them broadly. Automated patching systems can significantly streamline this process, reducing the risk of human error and ensuring timely updates. Remember to regularly scan for vulnerabilities using automated security scanners and promptly address any identified weaknesses. Proactive patching, coupled with vulnerability scanning, forms the bedrock of a secure virtual machine environment.

Next, consider the importance of strong passwords and multi-factor authentication (MFA). Weak passwords are a primary attack vector. Enforce strong password policies across all VMs and accounts, including regular password changes and complexity requirements. MFA adds an extra layer of security, requiring multiple forms of verification before granting access. Implement MFA wherever possible, especially for administrative accounts and access to critical systems. The combination of robust passwords and MFA significantly reduces the risk of unauthorized access, even if credentials are compromised. Remember to educate users about best practices.

Virtual machine isolation is crucial. Virtual machines, while isolated, aren't completely impervious to attacks. Proper configuration of your hypervisor and VMs ensures isolation. Utilize virtual networking technologies, such as VLANs, to segment your network and limit the impact of potential breaches. Regularly review the network configurations to ensure VMs are properly isolated and unauthorized access is prevented. Appropriate access control lists (ACLs) can further strengthen the security posture, limiting network traffic and preventing unauthorized communication between VMs. Effective isolation significantly mitigates the risk of a single compromised VM affecting others.

Intrusion detection and prevention systems (IDS/IPS) provide an additional layer of security. These systems monitor network traffic for malicious activity, alerting you to potential attacks and automatically blocking them. Deploy IDS/IPS solutions both within your virtual environment and at the network perimeter. Configure your IDS/IPS to detect and respond to various threats, including malware, port scans, and denial-of-service attacks. Regularly review the IDS/IPS logs to monitor system activity and identify potential threats. Proactive monitoring is crucial to ensure the effectiveness of your security measures.

Regular backups are essential. Data loss can have severe consequences. Implement a robust backup and recovery strategy, regularly backing up your VMs to a separate location. Consider using different backup methods, such as image-based backups and incremental backups, to ensure data redundancy. Regularly test your backup and recovery processes to ensure they function correctly and that you can restore your VMs in case of failure or attack. A well-defined backup and recovery plan minimizes the impact of data loss, enabling rapid recovery and minimal downtime.

Virtual machine security isn't a one-time task, but an ongoing process. Security measures need to be continuously assessed and updated. Regularly review and refine your security policies to stay ahead of emerging threats. Stay up-to-date with the latest security vulnerabilities and advisories, and adjust your security strategies accordingly. Regular security audits can help identify potential vulnerabilities and ensure compliance with security standards. Ongoing maintenance and vigilance are crucial to maintain the security of your virtual machines.

Security in a virtualized environment requires a multi-layered approach. Combining strong passwords, MFA, regular patching, VM isolation, IDS/IPS, and backup/recovery processes provides a robust defense against attacks. Continuous monitoring, assessment, and updating of your security strategy ensures ongoing protection. Remember, a proactive and well-maintained security posture is crucial to preventing security breaches and safeguarding your valuable data within your virtualized infrastructure. Prioritizing security is not just a best practice—it's a necessity in today's threat landscape. By implementing these security measures, you can significantly improve the security and resilience of your virtualized environment, protecting your valuable data and minimizing the risk of costly disruptions. Regular testing and review of these practices are paramount to ensure their continuing effectiveness and adaptation to ever-evolving threats. Remember that the cost of a security breach often significantly outweighs the investment in proactive security measures.

7.2 Security Best Practices

Regular patching is paramount. Keep your hypervisor, guest operating systems, and all applications updated with the latest security patches. Vulnerabilities are constantly discovered, and timely patching prevents exploitation. Neglecting this crucial step leaves your virtualized environment exposed to significant risks, potentially leading to data breaches or system failures. Prioritize critical patches immediately and schedule regular updates for less critical ones. A robust patching strategy, carefully planned and implemented, is the cornerstone of a secure virtualized infrastructure. Automate this process as much as possible to reduce the manual effort and human error.

Implementing strong access controls is absolutely vital. Employ robust authentication mechanisms such as multi-factor authentication (MFA) to restrict access to your virtualized environment. Limit administrative privileges to only those who absolutely need them, following the principle of least privilege. Regularly review and audit access rights to identify and revoke any unnecessary or outdated permissions. This prevents unauthorized access and mitigates the risk of insider threats. Consistent monitoring is key - track login attempts, privilege escalations, and other suspicious activities to promptly detect and respond to any security incidents.

Virtual machine isolation is a critical component. Employ strong isolation techniques to separate VMs from each other and from the underlying hypervisor. This prevents malicious code or vulnerabilities in one VM from affecting others. Consider using dedicated virtual networks for specific applications or groups of VMs. Properly configured firewalls at both the hypervisor and VM levels will further enhance security by limiting network access based on predefined rules. Regularly review and update these rules as your environment evolves to address emerging threats and operational needs.

Network security requires continuous vigilance. Segment your network using VLANs or VXLANs to isolate sensitive applications and data. Regularly review your network configurations to identify and mitigate potential vulnerabilities. Monitor network traffic for suspicious activity using intrusion detection and prevention systems. Employ robust firewall rules to control traffic flow and prevent unauthorized access. By diligently securing your network infrastructure, you create a significant barrier against malicious activity, protecting both your VMs and the data they contain.

Regular backups are essential for business continuity and disaster recovery. Implement a robust backup and recovery strategy that includes both virtual machine images and the underlying virtual infrastructure. Regularly test your backups to ensure they are recoverable. Consider using different backup locations, such as cloud storage or offsite tape backups, to protect against data loss in the event of a disaster. A comprehensive, well-tested backup strategy is crucial for minimizing disruption in the event of an unforeseen event. Remember that a poorly planned backup strategy is just as useful as having no backup strategy at all.

Regular security assessments are needed. Conduct regular security assessments, including penetration testing and vulnerability scans, to identify and mitigate potential threats. Use automated tools to scan your virtualized environment for known vulnerabilities and to assist with security assessment. These regular assessments will help identify previously unknown vulnerabilities. Engage experienced security professionals to assist with more complex or high-risk scenarios. Address identified vulnerabilities promptly. Implement robust monitoring and logging to track system events and potential security incidents.

Data encryption is crucial for protecting sensitive information. Encrypt all sensitive data both in transit and at rest. Utilize strong encryption algorithms and ensure proper key management. This prevents unauthorized access even if your system is compromised. Implement encryption at multiple levels, including the hypervisor, virtual machines, and individual applications. Remember to properly manage your encryption keys. This is essential to maintain the integrity and confidentiality of your encrypted data.

www.ingramcontent.com/pod-product-compliance
Lightning Source LLC
LaVergne TN
LVHW022015200926
848481LV00007B/2408
* 9 7 9 8 3 4 9 2 6 1 0 6 0 *